THE IM-PECK-ABLE GUIDE OF BIRDS

Carondelet High School
Ornithology Students

Copyright © 2020 Carondelet High School
All rights reserved.

All images are used with permission
and remain the property of their respective owners.

Carondelet High School
1133 Winton Drive
Concord, CA 94518
www.carondeleths.org

ISBN: 9798566043746

About the Authors

This book was written by the ornithology students of Carondelet High School, an all-girls Catholic institution located in Concord, California. This book not only represents the hard work and diligence of these young authors, but it also displays their dedication to the school's values of heart, faith, courage, and excellence. These authors are linked by the remarkable sisterhood that connects each student together to create one single body of ambitious and extraordinary young women.

Dedication

This book is dedicated to the Carondelet staff, students, and parents. We'd also like to thank the Lindsay Wildlife Experience and the Mount Diablo Audubon Society for their support.

Student Foreword

This book was created by the Class of 2021 seniors at Carondelet High School. Each student in the ornithology class with Mr. Buckles dedicated time towards researching a wide range of different species of birds in the Bay Area. Over the course of the semester, we put a lot of effort into researching our birds and synthesizing the information into this informative guide book. Thanks to Mr. Buckles and Mr. Walsh, we were able to set our goal in motion. This process went through multiple stages, from finding the perfect pictures, researching habitat and niches, and educating ourselves on bird preservation. We solidified ideas over the span of several class meetings and discussions, asking questions like: "What should the name of the book be?"; "Which areas should we focus on?"; and "What designs should we incorporate?" Mr. Walsh and Mr. Buckles allowed the seniors to take full reign on creating the book, while offering guidance. Students were broken off into separate teams, consisting of Lead Editors, Copy Editors, Layout, Publishing, Marketing, and Artwork. Through the hard work of each team, our book became a reality. Our goal is to help inspire people to get involved in the study of ornithology while raising awareness for the birds of the San Francisco East Bay.

Table of Contents

Chapter 1

Backyard Birds

Oak Titmouse

Baeolophus inornatus

The common name for this bird is Oak Titmouse, also referred to as Titmice. The scientific name is derived from *Baeolophus inornatus*. This creature is a passeriform, and in the family of paridae. The Size is smaller than a sparrow but bigger than a hummingbird.

The entire bird is not very colorful; it's mostly grey and black with blue and grey feet. There's a short grey crest on its head, with a white abdomen.

They have harsh calls but also are quite the song birds; their songs carry throughout the oak woods. They mate for life, and there is little to none sexual dimorphism; they look very much alike. The

Oak Titmouse breeding and wintering range is the bottom of Oregon, through most of California. The migration route is the Pacific Flyway.

The Oak Titmouse lives in open, and warm woods containing but not limited to pine, and oak trees. They rarely go outside of their comfort zone for territory. The Oak Titmouse has a wide variety of food it feasts on, including invertebrates, insects, acorns or seeds, and berries. They are foliage gleaners meaning that they get most of their food resources from barks and lichens of trees.

The Oak Titmouse is mostly hunted by snakes, mammals, and even large birds. It is a cavity bird, meaning that it lives in small holes, or natural crevices created in trees, preferably 40 feet above the ground, to stay away from ground predators.

Actions that can help improve the Oak Titmouse thrive is to have a protected designated area that isn't completely affected by urbanization, so that can help the Oak Titmouse revive the species. Humans can cause a huge impact on the Oak Titmouse species, because some of the biggest threats to this bird are urbanization, which can damage the bird's habitat, especially places to build their nests.

Brewers Blackbird

Euphagus cyanocephalus

The Brewer's Blackbird, or as it's known scientifically, *Euphagus cyanocephalus*, is a small, round bird. For reference, it could fit in the average person's hand. The males are black with blue undertones and the females are generally tawny brown. The differences between the sexes, however, largely stop here as Brewer's Blackbirds are sexually dimorphic in color and size.

These beautiful birds live near water and spend the majority of their time looking for food. Their preferred cuisine is seeds, nuts, and small insects. This diet does not make this bird a threat to other animals as they are herbivores, and their small size makes them a target for larger animals. The Brewer's Blackbirds have a quite distinctive walk; they bob their heads forward while walking on land. Both the male and female bird have two calls: a sharp trill and a low gurgle. The trill sounds metallic, and in this case sharp really means sharp. The call is incredibly piercing.

Calls, however, are not the only way these birds communicate. Like many other birds, Brewer's Blackbirds have a unique way of distinguishing themselves in order to find a mate. Their courtship is particularly interesting, and at times it is also aggressive. Size is generally an attractive feature in males, and it is no different for Brewer's Blackbirds. The male points his beak to the sky and tries to appear as large as possible by fluffing out his feathers, and spreading his wing and his tail.

Their eggs are generally a light green/grey color and speckled with brown. It takes about two weeks for an egg to hatch and two more for those chicks to leave the nest. Their nests are primarily found in bushes, and are made with small twigs, leaves, and sometimes even manure. Brewer's Blackbirds prefer open areas like fields, parks, and prairies. Their more natural habitat, open fields, have been slowly replaced by human made lawns. Moreover, these curious creatures have a wingspan of about 14.6 inches and fly by beating their wings quickly and powerfully. While some birds bob up and down in the air, Brewer's Blackbirds largely fly straight.

Black Phoebe

Sayornis nigricans

The Black Phoebe is a very small sized bird with an upright posture. They originated from southwest Oregon and California south which goes all the way through Central and South America. Their colors tend to range from black on the head, dark grey on the upper part of the body, chest, and wing feathers with a white belly. The behavior pattern from the Black Phoebe is sharp and high pitched whistle calls and laying low over any water source.

Their habitat tends to be along creeks and ponds located in the Southwest. During the winter, they tend to be in similar habitats but lower elevations. The range of breeding is 300 to 2100m above sea level and they do not leave their breeding ranges entirely. Their numbers also increase for breeding and migrate southward in the winter. Their wingspan is 27-28 cm. Their flight pattern is a weak fluttering flight with shallow wing beats. They tend to live in open habitats which could be streams, canyons, farmyards, near water, etc.

Their feeding niche is during the day in the nest and then usually go hunting for insects in the evening/night time. Furthermore, their predators range from foxes, snakes, cats, etc. The requirements for nesting is an incubation period of 15-18 days and a nesting period of 18-21 days. Their egg size ranges from the length of 0.7-0.8 inches and width of 0.5-0.6 inches. The nest they create usually stay 3-10 feet above water or the ground. The female tends to create the nests which can take up to 1-3 weeks after the male shows different possible locations for the nests which the female decides on the location.

Finally, actions that could improve the habitat around our school and at Heather Farms park for the Black Phoebe bird is to keep areas with water around because that is where they tend to find their food. It is also important to keep trees, plants, and dirt around for the Black Phoebe to make nests, catch other food sources, and to live in a peaceful and safe environment that will cause them no harm or for them to go extinct. The impact humans have on this bird is that we can create climate change which could cause them to not migrate at all and for them to die off.

House Finch

Haemorhous mexicanus

The House Finch, or scientifically known as the *Haemorhous mexicanus,* is a round, plump, and small size bird. Its plumage is either a dark dusty brown, or a mixture of greys. This small bird also features stripes of white feathers across its belly and down along the tail, as well as patches of dark brown or dark grey. Its chest and the front of its face is a bright red color, in which the eyes have a ring of grey feathers. The wings of the House Finch are short on the shorter side, making its tail quite long in comparison. They have grey beaks, and their feet are black, which are short in length. Also short in length would be their beaks, which aids them in tasks such as eating seeds off the ground.

A House Finches diet includes mostly vegetable matter, including seeds, berries, occasionally bits of various flowers, and small fruits in the late summer and into the fall season. These gregarious birds tend to gather at feeders and perch high up in the branches of trees. When perched, they tend to feed on the ground or on weed stalks. These birds move fairly slow on ground , and sit very still. Their flight is quite bouncy, which is very typical of most finches. House finches can be found in city parks, backyards, farms, and forests. The calls of the House Finch are light and sound almost squeaky. House Finches begin to couple off during winter within their flocks, and then remain together for the rest of the year. They then nest in a variety of places, such as the ivy of buildings, palms, cactus, and holes in manmade structures.

Their breeding habitat is overall urban and suburban across North America. Their nests are made around 12-15 feet above the ground. House Finches mostly reside permanently in the West, but some move to lower elevations if not already there for the winter. Those in the East are either permanent residents or migrate very long distances to the south in the fall. Most House Finches in the northeast United States migrate to the southern U.S. to stay for the winter. In the East, the female birds migrate farther south that the males do. House Finches in the east can stay in relatively the same area unless they are in high elevations, then they move to lower ones in the south.

California Towhee

Melozone crissalis

The *Melozone crissalis*, better known as the California Towhee, is your average backyard bird. The California Towhee is a medium sized bird with a chubby body and a short nape. Plumage appears to be grey/brown with hints of orange around the beak and undertail. While bird watching, you will be able to identify a California Towhee by its thick beak, hints of orange coloring, long tail, and clear metallic chirp.

These birds eat mostly seeds from different kinds of grasses and herbs, along with insects (mostly beetles and grasshoppers). California Towhees also eat berries such as elderberry, coffeeberry, and poison oak, acorns, and garden produce. Spiders, millipedes, and snails are also on the menu. Predators of the California Towhee could be hawks, foxes, or feral cats.

While looking for seeds and insects, this bird hops around on the ground as a common behavioral pattern. The California Towhee perches on shrubs, rooftops, or branches for long periods of time. This bird is commonly found in suburban areas, usually staying along the West Coast. They do not migrate during the year or have any other alternate breeding grounds. California Towhees can be found between Southern Oregon all the way down to the tip of Mexico.

The California Towhee's habitat name is known as a scrub. Their scrubs line the coastal slopes and foothills of the West Coast. Their habitats can also be located along streams and canyon bottoms next to desert slopes. They can also be spotted in shrubs in the suburbs of California neighborhoods! The nests of a California Towhee are found in shrubs or small trees. The nests are made by the female bird while the male observes. The outer cup is made from twigs, grasses, and dried flowers that she then lines with animal hair, sagebrush dark, and downy seeds.

Creating more space for California Towhees to live is essential for their lifestyle. Although they are at a low concern of conservation, if our state continues to build new buildings, these birds will lose their natural habitats. Even though California Towhees can make room in a suburban setting, it is still significantly important that they have real land to resort to, not just a backyard.

American Crow

Corvus brachyrhynchos

The American Crow has a thick-neck and a heavy, straight bill. In flight, these black birds have fairly broad wings and short tails that are rounded near the end. It has a thick-neck and a heavy, straight bill. Its call includes the caw, juvenile nasal call, rattle, crackoh & bell.

Crows breeding in upstate New York are partially migratory. Breeding pairs visit their breeding territory every year. Non-breeders will either stay in their original territory or spend time away. If some wander around the local area or join different foraging flocks, they may not visit their home. Other non-breeders leave the area entirely for several months. Crows tend to migrate out of the northernmost areas of their range. They can be seen crossing the Great Lakes in Spring and Fall, while migrating to Nebraska, Kansas and Oklahoma during winter. Crows in the southern range appear to be resident and nonmigratory. American Crows tend to live in close proximity to humans and other animals.

Their diet includes a wide range of foods including: seeds, nuts, grains, fruits, berries and small animals. They enjoy insects, aquatic fish, young turtles, crayfish, mussels, clams, eggs and nestlings, eating carrion and garbage. This wide-ranged diet makes survival easier.

American Crows find shelter easily, perching and building nests close to a tree trunk, as they are skilled at obtaining resources while utilizing their intelligence. Both members of a breeding pair help build the nest for crow nestlings, sometimes even young birds. The nests are made mostly of twigs and pine needles, weeds, soft bark, or animal hair. The nest size typically ranges from 6-19 inches across, with an inner cup about 6-14 inches across and 4-15 inches deep.

Predators of American Crows are hawks and owls. Hawks hunt crows during the day, and owls hunt them at night while in their roosts. Crows are able to live in any open place with a few trees to perch and a reliable food source, residing in both natural and human-made habitats.

California Scrub-Jay

Aphelocoma californica

The *Aphelocoma californica*, often known as the California Scrub-Jay, are large songbirds who have long and floppy tails. These birds often have a hunched-over posture and a straight bill with a hook at the tip. California Scrub-Jays are a blue and gray, with a clean, pale underside. They also have a blue necklace on their neck. As for their behavior patterns, these birds are quite assertive, vocal, slow, and glide through the air. The California Scrub-Jay lives mostly in the oak woodlands on the West Coast, along with living in pastures and orchards. Their field markings are blue and gray, with a pale underside and a blue necklace.

Like many other birds, the California Scrub-Jays call is a soft medley that can last up to 5 minutes. During the breeding and winter seasons, the California Scrub-Jay ranges from southern British Columbia and western Nevada. These birds migrate throughout parks, neighborhoods, and riverside woods near the Pacific Coast. Their habitat normally consists of scrubs, oak woodlands, and suburban yards. In addition, the California Scrub-Jay is a part of the omnivore feeding niches. Some predators include racoons, snakes, and crows. Due to their predators, Male or female birds often choose the sight that is normally low (6-14 inches high). Their nests are made of twigs with fine strands of plant fibers, and livestock hair. The nests take about 10 days to build and are about 6inches across.

California Scrub-Jays eat lots of insects, seeds, nuts, and fruit. In hopes of conserving our environment for these birds, there are several actions that could help improve the habitat around our school or at Heather Farms park. One thing to conserve the environment would be to leave the shrubs/small trees and possibly leave sunflower seeds and peanut feeders around.

Additionally, humans are affecting this species through climate change and pollution. With temperatures rising, the habitat of the California Scrub-Jay is ever-changing. For instance, fires are burning down the California Scrub-Jays habitat. Pollution from factories is also a key factor in the change of the birds habitat. Another way that humans are affecting these species is with all of the building and demolition of nature. Less trees lead to less nesting space for the California Scrub-Jay.

Anna’s Hummingbird

Calpyte anna

Anna's Hummingbird is also known as *Calypte anna.* Although on the smaller range, Anna’s are considered stocky and mid-sized for the hummingbird community. Anna’s Hummingbirds have bills that stick straight out and green and gray coloring on their bodies. Male Anna’s Hummingbirds have pink feathers covering their heads and throats while females are predominantly green and gray colored with no distinct brightly colored feathers. Anna’s Hummingbirds move rapidly and are difficult to spot when they are in motion. They usually hover over flowers or hummingbird feeders in search of nectar and sustenance. They have a distinctive, metallic song that is considered long for that of a hummingbird. Their songs can be described as a humming, along with more whistle-chipped notes.

Anna's Hummingbirds have large breeding grounds that are in various open habitats and gardens. These open areas range from the Pacific coast from around Vancouver south to northern Baja California as well as in southern Arizona. During the winter, breeding occurs in parts of southern Alaska/British Columbia. Anna's Hummingbirds either do not migrate or simply migrate very short distances for better feeding grounds.

Nesting for Anna's Hummingbirds begins in December. After males and females have their courtship display, nests sites are available on tree branches, shrubs, vines, on wires, or under eaves. Anna's Hummingbirds are known for their rapid flapping patterns, which can be observed while they hover over flowers and hummingbird feeders as they eat. They are found in gardens, chaparral, and open woodlands. Anna's Hummingbirds mostly feed off of nectar and insects, taking the nectar from flowers and feeding on tiny insects from the air. They hover while feeding at flowers, while extending its bill and tongue deep in the center of the flowers.

Anna's Hummingbirds live close to humans in many of their habitats and regions. Anna's Hummingbirds face big conservation threats, especially because of their tiny size possibly affecting their ability to adapt. Urbanization and infrastructure such as skyscrapers and taller buildings are harmful to Anna's Hummingbirds.

Chapter 2

Open Spaces Birds

White-breasted Nuthatch

Sitta carolinensis

The White-breasted Nuthatch is identified by its white belly and its white to grey to black ombré feathers. Along with their black crown. When looking for a White-breasted Nuthatch you will search for a small body no bigger than an apple. Their habitats include Forests, woodlots, groves, and shade trees especially ones along rivers, roads and public parks.

The White-breasted Nuthatch is located in habitats of woods and areas with Lots of surrounding trees. They can be found in most open areas. More specifically, the species can be found around woodland edges, parks, wooded suburbs, and yards. In these areas the White-breasted Nuthatch feeds on different types of foods and protects itself and their young from predators. Nuthatches feed off of multiple insects including larvae, wood-boring beetle larvae, other beetles, treehoppers, ants, caterpillars, click beetles, as well as some spiders and moths.

While it's helpful for the species to have such a wide variety of food, there are competitors and predators also searching for these foods. Predators of these birds include owls and hawks. White-breasted Nuthatches nest themselves in large natural tree cavities or in old woodpecker holes. They are usually 15-60' above ground. Females build nests in tree cavities, using a simple cup of bark fibers, grasses, twigs, hair.

The White-breasted Nuthatch is normally territorial throughout the year. They migrate and can be found all over North America. The breeding habitat of the White-breasted Nuthatch is made up of the woods across North America. It is continued into southern Canada to northern Florida and southern Mexico. The greatest threat to the White-breasted Nuthatch species is Tree companies. More specifically, logging. Logging and management practices that remove dead trees remove the species habitats and homes. Pushing out these birds not only has a negative impact on the species, but on the environment and humans. White-breasted Nuthatches have been found to be important when it comes to seed dispersal and germination in forests.

Red-tailed Hawk

Buteo jamaicensis

The Red-tailed Hawk, scientifically known as the *Buteo jamaicensis* is one of the most popular hawks in northern California. These birds have broad, rounded wings. Their wingspan can range from 45 to 52 inches in males, and 32 to 52 inches in females. They vary in color but are usually dark brown with cinnamon-red and white colorings. The female Red-tailed Hawk is often smaller and darker than the male. These birds' habitats are in woodlands, prairie groves, mountains, plains, and roadsides. In these places, they usually nest close to the ground. These hawks feed on small mammals, birds, and reptiles. They lay about 1-5 eggs, having a width of about 2 inches, and a length of 2.2 to 2.7 inches. Both partners will build the nest, and will incubate for 28 to 35 days. Red tailed-Hawks' eggs tend to be a buff white with occasional specks of brown or purple.

Despite some belief, Red-tailed Hawks can help some farmers by eating rodents and reptiles that could disturb their crops. This not only helps farmers, but also our ecosystem in controlling the populations of these small rodents, and providing homes for small animals. Their diet depends on the season or whatever they are able to get. You can find them all around North America, Canada, and even Alaska. Eastern birds tend to have more red colorings, while Red-tailed Hawks in the west are generally darker. Since their population is constantly thriving, you can easily spot them in your backyard or around your house. Red tailed-Hawks will stay with their same partner. During mating seasons, these birds can perform various rituals, such as interlocking talons and enticing the female birds with prey.

While Red-tailed Hawks are steadily increasing now, the majority of the deaths are caused by human interactions and creations. Shootings, automobile accidents, lead poisoning, and interference with nesting are the main causes of death. Wildfires and intense heat waves endanger both adults and young birds in their nests. It is important to do the best we can to make sure these birds keep their steady population. Red-tailed Hawks are extremely important to our environment, and without them, we pose risks of overpopulation of rodents and other small animals.

Turkey Vulture

Cathartes aura

Turkey Vultures are bigger than other raptors but smaller than eagles and condors. They are large dark birds with long broad wings, they have long "fingers" at their wingtips and long tails that extend past their toe tips in flight, up close they are dark brown with a featherless redhead. Look for them gliding relatively low to the ground, sniffing for carrion, or riding thermals up to higher vantage points. They may soar in small groups, roost in larger numbers, or see them on the ground in small groups, huddled around roadkill or dumpsters.

These birds range from southern Canada to the southernmost of South America. It inhabits a variety of open and semi-open areas. At night, they roost in trees, on rocks, and other secluded spots. In North America, these birds are more migratory in the West than in the East, and spend the winter in Central and South America. Turkey Vultures soar and glide extensively on thermals and mountain updrafts while migrating. When the currents become strong, the wings may be angled back, and frequently the wings are flexed below the horizontal until they are bow-shaped in outline. When they do need to flap their wings, their wingbeats are measured and deep.

As a part of their nesting "formation," several birds gather in a circle on the ground, and perform ritualized hopping movements around in a circle with wings partly spread. In the air, one bird may closely follow another, the two birds flapping and diving. Their nest sites are in sheltered areas, such as inside hollow trees or logs, in crevices in cliffs, under rocks, in caves, inside dense thickets, or in old buildings. Little or no nest built, eggs laid on debris or on the flat bottom of the nest site.

The East Contra Costa County Habitat Conservancy keeps the habitat lands clean and unbothered, and fine people who disturb the birds. The fees provide funding sources to purchase habitat lands or easements from willing sellers. Collected funds are also used for monitoring and any habitat enhancement or management actions.

Dark-eyed Junco

Junco hyemalis

These medium-sized sparrows can be about 5.5-6.3 inches in length, and weight up to 1.1 ounces! Even though the colors of these birds vary depending on where you are in the world, they are relatively the same.You can find these Juncos in the deep trees of forests, fields, and woodlands of the western US, the Appalachians, and right outside your door! Being small might be a struggle sometimes, it is also a good thing because these birds are foragers. Just like the name of these birds, their songs and calls are also very interesting. When you hear their calls/songs they are very high pitched almost like screeching but still peaceful. It sounds more like a "tweet tweet" than a "caw caw." It is not as bold as a crow, but smoother and calmer with a little bit of boldness.

The Dark-eyed Junco migrates in different directions. For

example, Juncos that breed in Canada or Alaska may migrate to the Southern United States for the winter, while Juncos that breed in the Rocky Mountains migrate a short distance or none at all. Deep inside these forests of douglas-fir, spruce, aspen, cottonwood, oak, maple, and hickory, many Juncos can be found. Luckily for these birds, they can find their food right inside of their habitats. With a diet of seeds from chickweed, buckwheat, lamb's quarter, and sorrel, it is quite easy to find something nice to snack on. However, while looking for food they do have to be cautious of their predators. Some possible predators are Sharp-Shinned Hawks, owls, and feral and domestic cats. When a predator approaches them, they will hide under anything they can find nearby. However, when a predator approaches their eggs and nest, a parent will call out, fly around, and dive at predators near the nesting area.

Just like specific habitat areas, each species also has nesting requirements and methods. For this species, the females choose and build the nest and its site. Most of the nests are found in a depression or other niches on sloping grounds, rock faces, or in the tangling of roots of an upturned tree. If a family of Juncos is near people, then they may nest under a building. Most nests are made out of fine lining of grasses, pine needles, twigs, and leaves and moss. From there, they are lined with grasses, ferns, rootlets, hair, and fine pieces of moss. The nests take about 3-7 days to build, and are about 3-5.5 inches in diameter, with an inner diameter of 2.4-2.8 inches, and depth of 1.6-2.8 inches.

Steller’s Jay

Cyanocitta stelleri

The Steller's Jay or *Cyanocitta stelleri,* has coloring similar to the blue jay; they boast a primarily blue body, but its head and upper body are black. Their bills and legs are black. They have a black crest of feathers on top of them . Most of them have white streaks on their foreheads and chins. They are around 12 inches long and can weigh anywhere from 100-140 grams. The males and females look the same. Forging is a practice they utilize if fresh food is not readily available. They travel in groups and are extremely social and vocal. They give a loud and repeated “shook shook shook” call while they fly, are perched, or in a state of aggression. This call may be heard during nesting season. These birds are usually quiet, but if threatened during nesting season, they may become aggressive. During this season, the male feeds the female. They have clutch sizes of 2-6 eggs.

The Steller's Jay is commonly found in western North America, ranging from the pacific coast to Alaska. They are commonly found in forested areas, but may also be spotted in residential areas as well. They are generally a resident species, but they do migrate a little during the spring and fall. They fly with a few flaps and generally glide and they beat their wings to navigate around the forests they live in. Their wings are shorter and rounded, as to make it easier for them to be among the trees. Their wingspan is around 17 inches.

They are omnivores, eating insects, seeds, berries, nuts, small animals, eggs. They will steal food from other birds or take food from people. Hawks, larger birds, coyotes, snakes, humans are their predators. To make a nest, stems, leaves, moss, are held together by mud. The inside is lined with pine needles, animal hair, or soft roots. They are around 2.5-3.5 inches deep and 6-7 inches tall.

The Steller's Jay relies on trees as its habitat and nesting site. Deforestation threatens the homes of many birds. These types of habitats should be placed under protection from deforestation in order to maintain the population of Steller's Jays and other species of birds. Despite this, the population of these birds is stable and they are not in any danger. They have large numbers, but if humans were to interfere with their habitat, they may become endangered.

Chapter 3

Water Birds

Green Heron

Butorides virescens

The scientific name for the Green Heron is *Butorides virescens*. The Green Heron has an oval shape with a long neck that often remains hidden, but can protrude out. They are mostly a dark green shade with a brown breast and neck. They have some white on their chest, orange legs, a black bill, and the underside of their wings is dark grey. Females are smaller and have muted color tones. During breeding season, males court females by calling loudly and stretching out their necks. Their songs are unique: it is a loud sound, similar to the sound of a dog barking. Their calls are raspy and can be brief or longer.

The male picks the location of the nest, which is often in a secluded area concealed by branches above or on water. The nests are 8-12 inches across and two inches deep. Green Herons live year round in Florida, the Caribbean, and the coasts of California and Mexico. During migration, they reside in the Southwest United States. During breeding season, they are on the coasts of Washington and Oregon and in the middle and eastern parts of the United States. In late winter and early spring, they begin migrating north, and return south in late August to October. Green Herons on the west coast migrate into Mexico, and Green Herons on the east coast migrate into Florida or the Gulf Coast. Their wingspan is 26 inches.

The wing load of the Green Herons is 0.35. The Green Heron can live in most areas as long as there is water and bushes nearby. They are common by the coast, ponds, lakes, marshlands, and swamps. They mostly consume small animals, small fish, and insects. Their eggs are eaten by grackles, crows, and snakes, while nestlings are consumed by racoons. Adults are sometimes hunted by birds of prey.

Green Herons often hurt aquaculture facilities and recreational fishing waters by eating a lot of the fish held there. They face risks from increasing development which disturbs their habitat and breeding areas. The Contra Costa Habitat Improvement Plan provides funding to protect habitats like the wetlands which Green Herons occupy. To improve their habitat, more trees and bushes should be planted close to shorelines so the Green Herons have more shrubbery to utilize near the water.

Great Blue Heron

Ardea herodias

Ardea herodias, also known as the Great Blue Heron, is the largest of the herons in North America. This large bird stands at around 4 feet tall and can weigh around 5 to 7 pounds. The Great Blue Heron, true to its name is a bluish grey color with a long white neck. The plume of the bird tends to be a dark blue/grey color as well. These birds tend to stand in shallow water areas and stand very still; while at times they have been seen wading slowly through the water. These are carnivorous birds, this entails that they eat smaller animals. The Great Blue Heron eats small fish in fresh and salt water bodies. They stay very still and try not to disrupt the water around them so that they can find fish near them. Nesting near the water allows the bird to be close to its primary food source at all times.

While the Great Blue Heron is a predator to fish, they are preyed on by other animals as well; Raccoons, Hawks and Eagles are known to prey on chick and adult Blue Heron. Crows and Raccoons are also known to eat the eggs of the bird. The Great Blue Heron inhabits Temperate Seasonal forest, Woodland, and Temperate grassland. This includes areas from coastal Alaska all the way down through North America, they have even been found as far south as the Caribbean. While the birds have year-round habitats, they tend to breed in the central north like Ontario,CA and North Dakota then migrate south later in the year. The Great Blue Heron makes very distinct calls, especially in its breeding grounds where the bird is most vocal.

While the Great Blue Heron has many habitats, it still needs protecting. The bird lives near the water sources where they get many resources. It is our job to make sure these waterways stay natural and unpolluted. It is also important that these rivers are not diverted in any way in order to keep these places habitable for the Great Blue Heron.

Mallard

Anas platyrhunchos

Mallards, or *Anas platyrhynchos*, are ducks with hefty bodies, rounded heads, and wide, flat bills. Males have a green head and neck with a brown body and a distinctive blue marking on their wings. Females are a light brown color throughout the whole body. The male birds have special colorful plumage during the spring and summer breeding season that helps them attract females. During breeding season, females and males can become aggressive.

Habitat types include fresh and salt-water wetlands like parks, ponds, rivers and lakes. They prefer calm, shallow sanctuaries, but can be found floating or surrounded by many bodies of freshwater. Their feeding niche includes algae, plants and insects. Their nests must be close to water and low bushes and long grass. The nest is mostly made out of grass and weeds. They are normally very territorial. Mallards use their calls when they feel threatened or when they are calling other birds. Females have a loud call that gets quieter over the span of their 2-10 quacks. Males give an even quieter, rasping, one or two-noted call. Red foxes are the most common predators among mallards.

They do not begin their migration until fall, around August or September. Mallards occur year-round across much of the United States. Mallards migrate across Canada and Alaska during the breeding season and travel across the Southern part of the United States and Mexico during the fall and winter season. They move between northern nesting grounds in the summer and warmer southern areas for resting in the winter. They need wetlands to feed and rest and San Francisco Bay is one of these habitats. Mallards have a low wing load due to their wings and body being proportional to each other. Their wingspan is around 2.7 to 3.2 feet for adults. They can beat their wings slowly while still remaining in flight.

Mallards create new reservoirs of plant biodiversity, benefiting other animals, the environment and even humans. Humans have negatively impacted Mallards, as they are popular to hunt and sell. They are also often hit by vehicles on roads. The Costa Costa Habitat Improvement plan fines people who harm these birds. The profits from the fines are used to protect the Mallards' habitat.

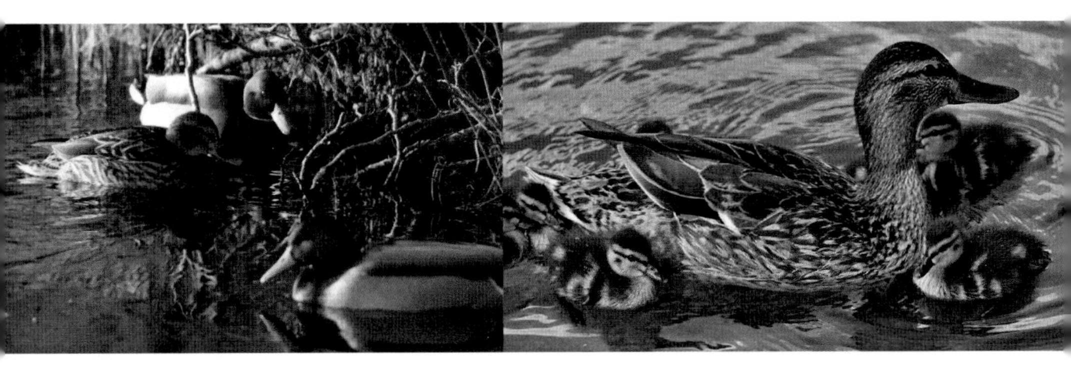

Black-necked Stilt

Himantopus mexicanus

The Black-necked Stilt, also known as the *Himantopus mexicanus*, is most likely to be spotted in habitats including shallow waters, mudflats, salt marshes, sewage ponds, evaporation pools, and flooded fields. This species is also negatively impacted by wetland destruction, which is a negative impact from humans' attempt to implement irrigation on the Black-necked Stilts' most prized habitats. Humans have also incited climate change to render the habitats of this bird by posing threats of wildfires which incinerate the Black-Necked Stilts habitat.

The Black-necked Stilt's behavior consists of mimicking injuries to impose a distraction. The Black-necked Stilt is smaller than the Willet and larger than the Lesser Yellowlegs. This species' external anatomy is primarily black with the exception of its white belly. Additionally, the Black-necked Stilt has a black crown, nape, back, and rump. This species also has long pink legs and a long thin black bill. As for songs, the Black-necked Stilt has no known song. Though, this species does possess a unique call, which is high-pitched and sometimes doubled. When the Black-necked Stilt becomes alarmed, it lets out calls in series.

The Black-necked Stilts migration is not limited to one consecutive location as this species winters in parts of California and the Gulf Coast and throughout Mexico and South and Central America. This species tends to breed in Washington from April to late September.

The Black-necked Stilt has a wingspan of 63-69 cm and a wing loading of .1334. The flight pattern of the Black-necked stilt consists of a direct straight level path and continuous flapping. The majority of their feeding prey is aquatic invertebrates, insects, and even small fish. As well, the main predators of this species include animals such as foxes, gulls, skunks, coyotes, and other birds. The Black-necked Stilts must nest on the ground. The location for this species nests is necessary to be located on surfaces above water, like small islands, clumps of vegetation, and even floating mats of alargar. This species will look for areas with soft sand or another substrate that can be pushed away to develop a nesting depression.

Bufflehead

Bucephala albeola

The Bufflehead (*Bucephala albeola*) is a small, plump duck that lives in coves along the atlantic and pacific ocean, inland ponds, and bodies of water, living throughout the United States during the winter, then breeding and migrating mainly in Canada. Buffleheads that breed west of the Rockies migrate to the Pacific Coast, while those that breed in central Canada migrate east or south. These ducks tend to be found bobbing in bodies of water while shallowly diving to catch small invertebrates.

The coloration of the Bufflehead is one of the most notable features of this species. The males have iridescent heads with large white patches running from the back of their eyes across the back of their head/ neck in an oval formation, and have a black back and white stomach plumage. The males also have a distinct thick white stripe on the top of their wings, which is very easy to spot when these birds are in flight. The females have dark brownish grey black body, and a distinct white patch under each eye. The Bufflehead has a 24 inch wingspan on average, and flies with fast rapid wing beats, while swaying from side to side while in flight.

The general habitat of the bufflehead is an area next to a body of water, this is important because this small duck strongly relies on aquatic creatures to feed on such as bug larvae, large zooplankton, shrimp, clams, and some small fish. Buffleheads have many predators, mainly predatory birds such as hawks, eagles, and falcons. They also have land predators such as squirrels, bears, and weasels. They nest in cavities which makes the females and young particularly vulnerable to squirrels and snakes. Their nests are usually made from other species of birds that burrow into the trees, mainly nests made by Northern Flickers and sometimes Pileated Woodpeckers. The females will lay their eggs in these holes and protect them from possible predators.

Buffleheads are not very vocal birds, the only calls they make tend to be females making a gutteral *cuk-cuk-cuk* sound when finding nesting sites/ protecting the area from other ducks. Other than that, males will usually only be vocal in their head bobbing courtship displays, making chattering noises in these times.

Great White Egret

Andrea alba

The Great White Egret *(Ardea alba)* is commonly found in the West Indies and southern Central America. They can be found in wetlands, streams, ponds, and tidal flats. During breeding season, they are found living in trees, shrubs, or anywhere other waterbirds can be found. Their chicks can be taken by racoons, Great Horned Owls, Red-tailed hawks, and other hawks, so males build nests to protect their young. Built from long sticks and twigs, their nests are elevated up to 100 feet off the ground. They eat mainly fish but can also be found eating other amphibians and invertebrates found along bodies of water.

Great White Egrets are slim, tall, and have a white body with long yellow legs, known for its "S-shaped" neck, green eyes, and bright yellow beak. They have an inconsistent call that is more quiet, almost trilling, croaky and nasally. Their song is very different from their call, having an almost wet, "blubbing" sound.

This bird commonly moves south for the winter, migrating during the day in small flocks. When it is breeding season, they can be found in southern Central America, and along the east coast. When they are not breeding, they can be found along the west coast and around South America. Their migration route typically ranges across the entire United States: the west coast, midwest, and the south.

Currently, Great White Egrets do not pose a threat on humans. However, humans pose a threat to them because in the late nineteenth and early twentieth centuries, 95% of them were hunted in North America. They were hunted for their plumes as people used them to make hats. In addition, people today contribute to the loss of the bird's habitat, with increased amounts of water and air pollutants. The Martinez Marina is surrounded by various chemical pollutants that can affect the bird's habitat if not removed. Hydrocarbons can also affect the bird's eggs and chicks as they can damage the life of its young.

The Contra Costa Habitat Improvement Plan provides conservation and developmental guidelines in order to protect the natural resources and habitat for the Great White Egret. This plan will also work to improve the permit process for species that are considered endangered and have wetland regulations.

Canadian Goose

Branta canadensis

The Canadian Goose is a large bird with a black head and neck and a brown and white body. Its posture stands straight up and is notably stiff. The bird is well known for its long neck and protruding chest that rounds down to a lower rounded belly. Canadian Geese are ground foragers; they eat from fields, roam around grass areas, and drink from ponds and shallow areas of water. They are well known for flocking together in groups and keeping with their mates for a year after breeding.

The Canadian Goose is primarily found in North America. For breeding season, Canadian Geese can be found in Canada from March to early April when they most typically breed. In the winter, however, the Geese tend to migrate south.

Canadian Geese live by marshes and areas where there are bodies of water nearby such as ponds, lakes and rivers. They get their food such as aquatic animals, grains, and grass in these places as well. For the Canadian Geese to nest they must have a partner which they begin to pair up at the age of three years old.

In their niche, they are prey to a number of predators that also reside in their habitat. The predators that target Canadian Geese are mainly raccoons, as well as skunks, foxes, coyotes, bobcats and crows to name a few. Raccoons prey on their goslings, or babies, and coyotes and bobcats tend to prey on adults.

The Canadian Goose can be spotted in local parks like Heather Farms in Walnut Creek and other regional parks in the Bay Area. They can also be spotted in Martinez Marina. As they reside in most parks that are heavily populated with people they encounter lots of trash. Due to human impact, their habitats are worsening and their marshes and vegetation are very much at risk.

Red-winged Blackbird

Agelaius phoeniceus

The Red-winged Blackbird, scientifically named *Agelaius phoeniceus* is a medium-sized bird. Their three primary colors are black, red, yellow, and sometimes orange. They naturally tend to stand somewhat with a tilt. They act as ground foragers. Their habitat consists of marshes, along watercourses, so they tend to nest in shrubs and eat insects. Their songs can be very loud and can be easily heard. The male's song sounds somewhat like a 'conk-la-lee'. They usually stay on high perches but females stay on the ground and weave through nests and find food. They usually live for about two years but the oldest reported Red-winged Blackbird to ever live died when it was fifteen years old, which is thirteen added years of life!

The Red-winged Blackbirds can be seen all around the US

year-round. However, the northern birds will migrate around 800 miles during the winter months. The southern birds do not migrate at all during the winter months. During the breeding season, they will travel up north towards Canada. Their wingspan is approximately 12-16 inches. The Red-winged Blackbird is a direct flier with quick motions. During the flight, they do not glide unless on occasion but dart straight to their destination. During landing, they glide to safely make it on the ground.

The Red-winged Blackbird's habitat includes fresh and saltwater marshes. They tend to be around any wet areas, but in winter they can be seen in crop fields. Red-winged Blackbirds are very aggressive. They are very quick to attack something that may be seen as a threat. They are preyed on by raccoons, skunks, and snakes. They avoid predators during nesting by putting their nest 1 to 2 meters above water. Their nests range from 4-7 inches wide and 3-7 inches in length. Females choose the nest and help build it. It is mostly made up of wet leaves. She will pick the nest based on more of where it's located. Typically it is near the ground and water.

The Red-winged Blackbird is very helpful during the crop season. They help keep grains and seeds available for people. They keep their habitats clean and they have a safe place to nest, mate, and reproduce. Pollution is a huge problem that humans have made for these birds. Th Contra Costa Habitat Improvement Plan give them a chance to keep their habitat safe.

Glossary

Breeding season When birds reproduce; typically during spring.

Calls Short and simple sounds followed by gaps of silence made by birds. They repeat and can be used for announcing threats, begging, and to warn other birds.

Courting Before breeding, birds court each other. The male usually courts the female using specific tactics to draw the female's attention. This can be in the form of appearances, gifts, dancing, and more.

Endangered Species of birds at risk of extinction.

Feeding niche- What the bird's diet consists of in their specific environment.

Field markings Markings that can be used to identify a bird when spotted in the wild.

Fledgling A baby bird.

Foliage gleaners-Birds that get their food through plucking it from the foliage or finding it under rocks, houses, or other things.

Ground forager Birds that gather food by finding it on the ground.

Habitat The area/environment in which a bird lives.

Herbivore Only eats plants, not flesh of other animals.

Omnivore Eats both plants and animals.

Migration Period of time in which birds move from one place to another; which usually occurs annually.

Plumage A bird's feathers.

Predators Animals that hunt other animals for food.

Sexual dimorphism Differences in the appearance of males and females of the same species.

Songs Long and continuous sounds, more complicated than calls. They include a series of elements or phrases and can be used to attract mates or announce territory.

Urbanization Land used for human habitation, which destroys the natural environment.

Wing load Measurement that compares the body mass to wing area of a bird. ~ Formula: wing loading = body mass (kg)/wing area (m2)

Wingspan The length of a bird's wings from tip to tip, when completely stretched out.

Bibliography

Ch. 1: Backyard Birds

"Anna's Hummingbird." Hummingbird Facts and Information, 2014, www.hummingworlds.com/annas-hummingbird/.

"Black Phoebe Identification, All About Birds, Cornell Lab of Ornithology." , *All About Birds, Cornell Lab of Ornithology*, www.allaboutbirds.org/guide/Black_Phoebe/id.

"Black Phoebe." *Audubon*, 11 Oct. 2019, www.audubon.org/field-guide/bird/black-phoebe.

"Black Phoebe." *National Geographic*, 21 Sept. 2018, www.nationalgeographic.com/animals/birds/b/black-phoebe.

"Brewer's Blackbird Overview, All About Birds, Cornell Lab of Ornithology." *Overview, All About Birds, Cornell Lab of Ornithology*, Cornell University, www.allaboutbirds.org/guide/Brewers_Blackbird.

"California Towhee Overview, All About Birds, Cornell Lab of Ornithology." *Overview, All About Birds, Cornell Lab of Ornithology*, www.allaboutbirds.org/guide/California_Towhee.

"California Scrub-Jay." *Audubon*, 2 Mar. 2020, www.audubon.org/field-guide/bird/california-scrub-jay.

"California Scrub-Jay Identification, All About Birds, Cornell Lab of Ornithology." , *All About Birds, Cornell Lab of Ornithology*, www.allaboutbirds.org/guide/California_Scrub-Jay/id.

"Flyways." *U.S. Fish & Wildlife Service - Department of the Interior*, www.fws.gov/birds/management/flyways.php.

"House Finch." *Audubon*, 9 Dec. 2019, www.audubon.org/field-guide/bird/house-finch.

"House Finch Overview, All About Birds, Cornell Lab of Ornithology." *Overview, All About Birds, Cornell Lab of Ornithology*, www.allaboutbirds.org/guide/House_Finch/overview.

Kaufman, Kenn. "Anna's Hummingbird." Audubon, 12 Nov. 2019, www.audubon.org/field-guide/bird/annas-hummingbird.

Oak Titmouse Overview, All About Birds, Cornell Lab of Ornithology. www.allaboutbirds.org/guide/Oak_Titmouse

"Oak Titmouse." *Audubon*, 24 Dec. 2019, www.audubon.org/field-guide/bird/oak-titmouse.

University , Cornell. "Anna's Hummingbird Overview, All About Birds, Cornell Lab of Ornithology." Overview, All About Birds, Cornell Lab of Ornithology, 2019, www.allaboutbirds.org/guide/Annas_Hummingbird.

Walte, Mitch. "Anna's Hummingbird." Whatbird.com, 2013, identify.whatbird.com/obj/167/_/Annas_Hummingbird.

Ch. 2: Open Space Birds

"Dark-Eyed Junco Identification, All About Birds, Cornell Lab of Ornithology." , *All About Birds, Cornell Lab of Ornithology*, www.allaboutbirds.org/guide/Dark-eyed_Junco/id.

"Dark-Eyed Junco Life History, All About Birds, Cornell Lab of Ornithology." , *All About Birds, Cornell Lab of Ornithology*, www.allaboutbirds.org/guide/Dark-eyed_Junco/lifehistory.

"Dark-Eyed Junco Range Map, All About Birds, Cornell Lab of Ornithology." , *All About Birds, Cornell Lab of Ornithology*, www.allaboutbirds.org/guide/Dark-eyed_Junco/maps-range..

Dark-Eyed Junco(Slate-Colored) Song, LesleytheBirdNerd, 27 Apr. 2012, youtu.be/vlJUsAl4YCA.

"Dark-Eyed Junco." *Audubon*, 31 Mar. 2020, www.audubon.org/field-guide/bird/dark-eyed-junco.

"Dark-Eyed Junco: National Geographic." *Animals*, 24 Sept. 2018, www.nationalgeographic.com/animals/birds/d/dark-eyed-junc o.

"East Contra Costa County Habitat Conservancy: Overview/History." *East Contra Costa County Habitat Conservancy | Overview/History*, www.contracosta.ca.gov/depart/cd/water/HCP/overview.html.

"Learning." *Red-Tailed Hawk*, North Carolina Wildlife Resources Commission, www.ncwildlife.org/Learning/Species/Birds/Red-Tailed-Hawk.

Person. "Red-Tailed Hawk." *Hawkwatch International*, Hawkwatch International, 12 Aug. 2014, hawkwatch.org/learn/factsheets/item/104-redtailed-hawk.

"Red-Tailed Hawk Overview, All About Birds, Cornell Lab of Ornithology." *Overview, All About Birds, Cornell Lab of Ornithology*, www.allaboutbirds.org/guide/Red-tailed_Hawk/overview.

"Red-Tailed Hawk." *Audubon*, 18 June 2020, www.audubon.org/field-guide/bird/red-tailed-hawk.

"Red-Tailed Hawk." *National Geographic*, *https://www.nationalgeographic.com/animals/birds/r/red-tailed-hawk.*

"Steller's Jay." *Audubon*, 18 June 2020, www.audubon.org/field-guide/bird/stellers-jay.

"Steller's Jay Identification, All About Birds, Cornell Lab of Ornithology." *All About Birds, Cornell Lab of Ornithology*, www.allaboutbirds.org/guide/Stellers_Jay/id.

"Steller's Jay." *Steller's Jay - Facts, Diet, Habitat & Pictures on Animalia.bio*, animalia.bio/stellers-jay.

"Turkey Vulture Overview, All About Birds, Cornell Lab of Ornithology." *Overview, All About Birds, Cornell Lab of Ornithology*, www.allaboutbirds.org/guide/Turkey_Vulture.

"Turkey Vultures Are the Most Migratory of All Vultures." *Hawk Mountain Sanctuary*, www.hawkmountain.org/raptors/turkey-vulture.

"Turkey Vulture." *Audubon*, 26 Mar. 2020, www.audubon.org/field-guide/bird/turkey-vulture.

"White-Breasted Nuthatch." *Audubon*, 19 Feb. 2020, www.audubon.org/field-guide/bird/white-breasted-nuthatch

WINGMASTERS Species, www.wingmasters.net/tvulture.htm.

Ch. 3: Water Birds

"Black-Necked Stilt Overview, All About Birds, Cornell Lab of Ornithology." Overview, All About Birds, Cornell Lab of Ornithology, www.allaboutbirds.org/guide/Black-necked_Stilt/overview.

Bouglouan, Nicole. "Green Heron Butorides Virescens." *Green Heron*, www.oiseaux-birds.com/card-green-heron.html

"Bufflehead." *Audubon*, 18 Nov. 2019, www.audubon.org/field-guide/bird/bufflehead.

"Bufflehead." *Chesapeake Bay Program*, www.chesapeakebay.net/S=0/fieldguide/critter/bufflehead.

"Bufflehead Identification, All About Birds, Cornell Lab of Ornithology." *, All About Birds, Cornell Lab of Ornithology*, www.allaboutbirds.org/guide/Bufflehead/id.

"Conserving Natural Lands and Sustaining Economic Development." East Contra Costa County Habitat Conservation Plan Association, Oct. 2006.

Dewey, Tanya, and Josh Butzbaugh. "Butorides Virescens." *Animal Diversity Web*, University of Michigan, 2001, animaldiversity.org/accounts/Butorides_virescens.

Dewey, Tanya, and Josh Butzbaugh. "Green Heron ." *BioKIDS*, University of Michigan, 2001, www.biokids.umich.edu/critters/Butorides_virescens.

"East Contra Costa County Habitat Conservancy: Overview/History." *East Contra Costa County Habitat Conservancy | Overview/History*, www.contracosta.ca.gov/depart/cd/water/HCP/overview.html.

"Great Egret." *Sacramento Splash*, www.sacsplash.org/critter/great-egret.

"Green Heron ." *All About Birds, Cornell Lab of Ornithology*, Cornell University , www.allaboutbirds.org/guide/Green_Heron/overview.

"Green Heron." *Chesapeake Bay Program*, Chesapeake Bay Program, www.chesapeakebay.net/discover/field-guide/entry/green_he ron.

Hoy, Michael D. "Herons and Egrets." United States Department of Agriculture.

Huth, John. "Bucephala Albeola (Bufflehead)." *Animal Diversity Web*, animaldiversity.org/accounts/Bucephala_albeola.

Johnson, David H. "Wing Loading in 15 Species of North American Owls." North Central Research Station.

Jones, Jessica. "Ardea Alba (Great Egret)." *Animal Diversity Web*, animaldiversity.org/accounts/Ardea_alba.

Kushlan, J. A., M. J. Steinkamp, K. C. Parsons, J. Capp, M. A. Cruz, M. Coulter, I. Davidson, L. Dickson, N. Edelson, R. Elliott, R. M. Erwin, S. Hatch, S. Kress, R. Milko, S. Miller, K. Mills, R. Paul, R. Phillips, J. E. Saliva, W. Sydeman, J. Trapp, J. Wheeler and K. Wohl (2002). Waterbird conservation for the Americas: The North American waterbird conservation plan, version 1. Washington, DC, USA.

Lutmerding, J. A. and A. S. Love. (2019). Longevity records of North American birds. Version 1019 Patuxent Wildlife Research Center, Bird Banding Laboratory 2019.

McCrimmon Jr., Donald A., John C. Ogden and G. Thomas Bancroft. (2011). Great Egret (*Ardea alba*), version 2.0. In The Birds of North America (P. G. Rodewald, editor). Cornell Lab of Ornithology, Ithaca, New York, USA.

North American Bird Conservation Initiative. (2014). The State of the Birds 2014 Report. US Department of Interior, Washington, DC, USA.

Pacific, Aquarium of the. "Bufflehead." *Aquarium of the Pacific | Online Learning Center | Bufflehead*, www.aquariumofpacific.org/onlinelearningcenter/species/bufflehead.

"Red-Winged Blackbird Identification, All About Birds, Cornell Lab of Ornithology." , *All About Birds, Cornell Lab of Ornithology*, www.allaboutbirds.org/guide/Red-winged_Blackbird/id.

Sauer, J. R., J. E. Hines, J. E. Fallon, K. L. Pardieck, Jr. Ziolkowski, D. J. and W. A. Link. The North American Breeding Bird Survey, results and analysis 1966-2013 (Version 1.30.15). USGS Patuxent Wildlife Research Center (2014b). Available from www.mbr-pwrc.usgs.gov/bbs.

Sibley, D. A. (2014). The Sibley Guide to Birds, second edition. Alfred A. Knopf, New York, NY, USA.

Sturm, Colleen. "CREATURE FEATURE - Green Heron." *Friends of the Rouge*, 18 Sept. 2020, therouge.org/creature-feature-green-heron.

Photo Credits

All images are used with permission and
remain the property of their respective owners.

Ch. 1: Backyard Birds

Oak Titmouse

Becky Matsubara on Flickr
www.flickr.com/photos/beckymatsubara/40867098010

Becky Matsubara on Flickr
www.flickr.com/photos/beckymatsubara/39701419083

siamesepuppy (username) on Flickr
www.flickr.com/photos/siamesepuppy/32402753197

Brewer's Blackbird

Jennifer Beebe on Pixabay
cdn.pixabay.com/photo/2018/06/22/15/57/grackle-3491014_1280.jpg

Chris Briggs on Unsplash
images.unsplash.com/photo-1571160466604-d7a0def2a777

Dick Daniels on Wikimedia
upload.wikimedia.org/wikipedia/commons/5/5e/Brewer%27s_Blackb
ird_male_RWD4.jpg

Black Phoebe

PublicDomainImages on Pixabay
cdn.pixabay.com/photo/2014/07/08/13/58/nigricans-387335_960_7
20.jpg

Kramer Gary on Pixnio.com
pixnio.com/free-images/fauna-animals/birds/black-phoebe-passerine
-bird-sayornis-nigricans-ttyrant-flycatcher-family-361x544.jpg

Sarangib on Pixabay
cdn.pixabay.com/photo/2015/10/24/18/45/black-phoebe-1004804_
960_720.jpg

House Finch

Bryan Hanson on Unsplash
images.unsplash.com/photo-1590158868156-2ec74ffcfbaa

Jeremy Stanley on Unsplash
images.unsplash.com/photo-1557948649-8624c28a797e

GeorgiaLens on Pixabay
pixabay.com/photos/bird-wildlife-male-house-finch-3285162

California Towhee

Carl Thomson on Pexels
www.pexels.com/photo/bird-california-towhee-1209062

PublicDomainImages on Pixabay
cdn.pixabay.com/photo/2014/07/08/13/28/bird-387103_960_720.jpg

Dick Daniels on Wikimedia
upload.wikimedia.org/wikipedia/commons/thumb/7/7a/California_Towhee_RWD4.jpg/800px-California_Towhee_RWD4.jpg

American Crow

Tom Swinnen on Pexels
images.pexels.com/photos/946344/pexels-photo-946344.jpeg

Frank Cone on Pexels
www.pexels.com/photo/grayscale-photo-of-bird-perched-on-branch-2291884

Chris LeBoutillier
www.pexels.com/photo/black-bird-on-top-of-brown-driftwood-929384

California Scrub-Jay

Noah Boyer on Unsplash
images.unsplash.com/photo-1573839817670-18ec12bd1940

Robert Burton, USFWS on Pixnio
pixnio.com/free-images/fauna-animals/birds/jay-birds-pictures/western-scrub-jay-on-a-branch-aphelocoma-californica-362x544.jpg

Dulcey Lima on Unsplash
images.unsplash.com/photo-1601765927633-77a9befc1a76

Anna's Hummingbird

Bryan Hanson on Pixabay
cdn.pixabay.com/photo/2020/03/22/02/36/male-4955658_960_720.jpg

Kvnga on Unsplash
images.unsplash.com/photo-1596832294353-64d120a6cca1

Bryan Hanson on Unsplash
images.unsplash.com/photo-1551673099-bf9469cd3d3a

Ch. 2: Open Space Birds

White-breasted Nuthatch

Joan Tisdale
nas-national-prod.s3.amazonaws.com/aud_gbbc-2016_white-breasted-nuthatch_35889_kk_mi_photo-joan-tisdale_adult-male.jpg

Jack Blumer on Pixabay
cdn.pixabay.com/photo/2020/10/04/23/05/white-breasted-nuthatch-5627753__480.jpg

Dick Daniels on Wikimedia
en.wikipedia.org/wiki/White-breasted_nuthatch#/media/File:White-breasted_Nuthatch_RWD1.jpg

Red-tailed Hawk

edbo23 on Pixabay
pixabay.com/images/id-1827949

sdc140 on Pixabay
cdn.pixabay.com/photo/2018/12/13/21/48/red-tailed-hawk-3873818_640.jpg

Joshua J. Cotten on Unsplash
images.unsplash.com/photo-1601246733383-4ad305e42a23

Turkey Vulture

Greg Seymour on Unsplash
pixabay.com/photos/bird-vulture-turkey-vulture-animal-1033767/

cdrying on Unsplash
unsplash.com/photos/QKaK7s1nP7s

Karney Lee on Pixnio
pixnio.com/free-images/fauna-animals/birds/turkey-vulture-bird-small-unfeathered-head-and-hooked-bill-aid-these-scavenger-in-consuming-carrion-725x483.jpg

Dark-eyed Junco

John Duncan on Unsplash
unsplash.com/photos/aYa5CO7VKcE

Rita Starceski on Unsplash
unsplash.com/photos/nX7AQPba_R4

Daniel Bisett on Pixabay
pixabay.com/photos/dark-eyed-junco-bird-bird-in-hand-2145227

Steller's Jay

Pete Nuij on Unsplash
images.unsplash.com/photo-1594341638160-f94922eb1d5f

Veronika Andrew on Pixabay
cdn.pixabay.com/photo/2020/08/30/18/44/stellers-jay-5530337_960
_720.jpg

Benjamin Grant on Unsplash
unsplash.com/photos/FBWhCcqsdYs

Ch. 3: Water Birds

Green Heron

Joshua J. Cotten on Unsplash
images.unsplash.com/photo-1591200352480-8ba694f3abf1

Joshua J. Cotten on Unsplash
mages.unsplash.com/photo-1567467934383-2358903719c8

Charles Jackson on Unsplash
images.unsplash.com/photo-1592248438411-75f5aab2ce7b

Great Blue Heron

terrysartifacts on Pixabay
cdn.pixabay.com/photo/2015/04/28/19/55/great-blue-heron-744257
_1280.jpg

Dr Thomas Barnes on Pixnio
pixnio.com/fauna-animals/birds/heron-bird/great-herons/great-blue-heron-bird-ardea-herodias

wileydoc on Pixabay
cdn.pixabay.com/photo/2020/09/07/18/21/great-blue-heron-5552562_1280.jpg

Mallard

Capri23auto on Pixabay
pixabay.com/photos/duck-drake-water-bird-duck-bird-3648416

Anne Nygard on Unsplash
unsplash.com/photos/ku5z_F9p7_Q

alexa_fotos on Pixabay
pixabay.com/photos/mallard-chicks-baby-swim-small-2793927

Black-necked Stilt

Kabomani Tapir on Pixabay
pixabay.com/photos/black-neck-stilt-stilt-wader-4401359

Public Domain on Pixnio
pixnio.com/fauna-animals/birds/black-necked-stilt-himantopus-mexicanus

Lisa Tanner on Wikimedia
en.wikipedia.org/wiki/Black-necked_stilt#/media/File:Himantopus_mexicanus_-Morro_Bay,_California,_USA_-flying-8.jpg

Bufflehead

Mdf on Wikimedia
en.wikipedia.org/wiki/Bufflehead#/media/File:Bucephala-albeola-007
.jpg

Bryan Hanson from Pixabay
cdn.pixabay.com/photo/2020/03/05/00/12/duck-4903047_960_720.jpg

Bill Bouton on Wikimedia
en.wikipedia.org/wiki/Bufflehead#/media/File:Bucephala_albeola_-Sa
n_Luis_Obispo,_California,_USA_-flying-8.jpg

Great White Egret

Simerpreet Cheema on Unsplash
images.unsplash.com/photo-1601095364905-0af4a36b0665

Joshua J. Cotten on Unsplash
images.unsplash.com/photo-1597268265977-0eda631f5ce5

Firecloak on Unsplash
images.unsplash.com/photo-1513915018042-bacf14fdf8c5

Canada Goose

Alexas_Fotos on Pixabay
cdn.pixabay.com/photo/2018/09/12/22/14/canada-goose-3673347_
1280.jpg

Jeffery Hamilton on Unsplash
images.unsplash.com/photo-1585673431620-5c0f6f84d5e2

Mario Medeiro on Unsplash
images.unsplash.com/photo-1605476249213-d7a5d0722411

Red-winged Blackbird

Walter Siegmund on Wikimedia
en.wikipedia.org/wiki/Red-winged_blackbird#/media/File:Agelaius_phoeniceus_0110_taxo.jpg

Karney Lee on Pixnio
pixnio.com/fauna-animals/birds/blackbirds-pictures/red-winged-blackbird/female-red-winged-blackbird-bird

Public Domain on Pixnio
pixnio.com/fauna-animals/birds/blackbirds-pictures/red-winged-blackbird/red-winged-blackbird-close-up-agelaius-phoeniceus

Made in the USA
Middletown, DE
27 January 2021

32542483R00042